SCIENTIFIC AMERICAN EDUCATIONAL PUBLISHING

POWER UP!

10 FUN EXPERIMENTS ABOUT ENERGY

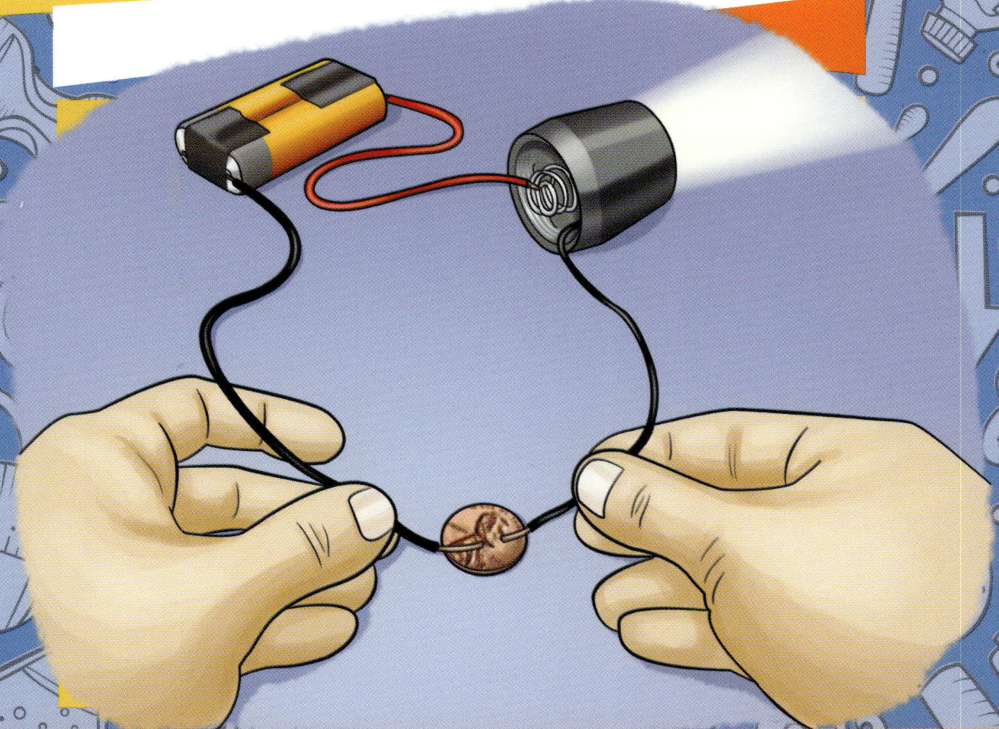

BRING SCIENCE HOME

Published in 2024 by The Rosen Publishing Group, Inc.
2544 Clinton Street, Buffalo, NY 14224

Contains material from Scientific American, a division of Springer Nature America, Inc., reprinted by permission as well as original material from The Rosen Publishing Group.

First Edition

Editor: Kristen Rajczak Nelson
Designer: Rachel Rising

Activity on p. 5 by Science Buddies (January 23, 2014); p. 11 by Science Buddies, (January 12, 2012); p. 15 by Exploratorium (February 19, 2015); p. 19 by Science Buddies (December 22, 2011); p. 23 by Science Buddies, Ben Finio (June 16, 2016); p. 29 Science Buddies (July 23, 2015); p. 37 by Science Buddies, Sabine De Brabandere (December 12, 2019); p. 45 by Exploratorium (March 19, 2015); p. 49 by Science Buddies (May 15, 2014); p. 55 by Science Buddies (September 17, 2015).

All illustrations by Continuum Content Solutions

Photo Credits: pp. 5, 11, 19, 23, 29, 37, 45, 49, 55 Anna Frajtova/Shutterstock.com; pp. 5, 9, 11, 13, 29, 34, 37, 41, 49, 52 cve iv/Shutterstock.com.

Library of Congress Cataloging-in-Publication Data
Names: Scientific American Educational Publishing, editor.
Title: Power up! : 10 fun experiments about energy / Scientific American Editors.
Description: Buffalo, NY : Scientific American Educational Publishing, an imprint of Rosen Publishing, [2024] | Series: Bring science home | Includes index.
Identifiers: LCCN 2023026536 (print) | LCCN 2023026537 (ebook) | ISBN 9781725350014 (library binding) | ISBN 9781725350007 (paperback) | ISBN 9781725350021 (ebook)
Subjects: LCSH: Force and energy--Experiments--Juvenile literature.
Classification: LCC QC73.4 .P73 2024 (print) | LCC QC73.4 (ebook) | DDC 531/.6--dc23/eng/20230815
LC record available at https://lccn.loc.gov/2023026536
LC ebook record available at https://lccn.loc.gov/2023026537

Manufactured in the United States of America

CPSIA Compliance Information: Batch #CWSA24. For further information contact Rosen Publishing, New York, New York at 1-800-237-9932.

CONTENTS

* THESE ACTIVITIES INCLUDE SCIENCE FAIR PROJECT IDEAS.

INTRODUCTION

When you think of energy, you may think about the electricity that powers your lights, cell phone, and other electronics. But energy is all around us, even in the food you eat! The following experiments will show you how we use and see energy in our day-to-day lives.

Projects marked with ⚛ include a section called Science Fair Project Ideas. These ideas can help you develop your own original science fair project. Science fair judges tend to reward creative thought and imagination, and it helps if you are really interested in your project. You will also need to follow the scientific method. See page 61 for more information about that.

Static Science

How Well Do Different Materials Make Static Electricity?

CREATE ELECTRICITY WITH COMMON HOUSEHOLD OBJECTS!

Have you ever noticed that some types of clothes are more susceptible to static cling than others? For example, a wooly sweater can have a lot of static cling, but clothing made out of cotton doesn't cling nearly as much. How well do other materials around the house produce static electricity? In this science activity, you'll explore this by making a simple, homemade electroscope (an instrument that detects electric charges) and testing it out. The results may shock you!

PROJECT TIME

45 to 60 minutes

KEY CONCEPTS

Electricity
Materials
Conductivity
Electrons

BACKGROUND

Static electricity is the buildup of electrical charge on an object. This charge can be suddenly discharged (such as when a lightning bolt flashes through the sky) or it can cause two objects to be attracted to one another. Socks fresh out of the dryer that cling together are a good example of this attraction in action. Specifically, static cling is an attraction between two objects with opposite electrical charges, one positive and one negative.

Static electricity can be created by rubbing one object against another object. This is because the rubbing releases negative charges, called electrons, which can build up on one object to produce a static charge. For example, when you shuffle your feet across a carpet, electrons can transfer onto you, building up a static charge on your skin. You can suddenly discharge the static charge as a shock when you touch a friend or some objects.

Whereas objects that have opposite charges are attracted to one another (such as clingy, freshly dried socks), objects that have the same charge repel. This principle is used in making an electroscope, which is a scientific instrument that detects electrical charges.

MATERIALS

- Styrofoam cup
- Sharp pencil or skewer
- Plastic drinking straw
- Aluminum pie pan
- Tape
- Clay (optional)
- Scissors
- Thread
- Aluminum foil
- Styrofoam plate (Alternatively, the Styrofoam lid from a takeout food container would work too.)
- Balloon
- A nonmetal desk or table (For example, a wooden, plastic, or glass desk or table would work.)
- At least one material to test (It should be no larger than the plate or can be folded to be small and able to lay flat. Some of the materials you could test include polyester, nylon, cotton, wool, silk, aluminum, plastic wrap, copper, wood, and tissue paper.)

PREPARATION

- To make your homemade electroscope, first make two holes near the bottom of a Styrofoam cup (on opposite sides of the cup), such as by pushing a sharp pencil or skewer through the cup. (Always use caution and an adult's help when manipulating sharp objects.) Push a plastic straw through both holes.

- Next, use tape or four small clay balls to secure the cup's opening to the aluminum pan (flipping the cup upside down). If you are using clay, stick four little balls of clay (each about 0.5 inch [1.3 cm] in diameter) to the rim of the cup, then turn the cup upside down and stick it to the aluminum pan. Adjust the straw's position so that one end of the straw is right above the edge of the pan.

- Cut a piece of thread with a length that's about two or three times the distance between the straw and the pan's edge. Tie a few knots in one end of the thread.

- Cut out a 1-inch (2.5-cm) square of aluminum foil. Use it to make a ball around the knots in the thread. The ball should be about the size of a marble or smaller and be just tight enough so it does not fall off the thread.

- Tape the thread to the tip of the straw so that the ball of foil hangs straight down from the straw, just touching the edge of the pan. *Why do you think it's important that the ball touches the pan?* Adjust the straw's position if needed. (If the end of the thread without the ball is dangling down and touching the pan, cut it so it does not touch the pan.)

- Your homemade electroscope is now ready for testing! When working with electricity, take precautions and beware of electric shock.

PROCEDURE

- To test your electroscope, create some static electricity by rubbing a blown-up balloon on a Styrofoam plate. Rub the Styrofoam plate several times with the balloon. *How do you think this creates static electricity?*
- Quickly place the electrically charged plate on a desk or table (that is not metal). Then place the electroscope you made on top of the plate. Be sure to only hold the electroscope by the Styrofoam cup and not by the aluminum pan, otherwise it will not work! *Why do you think this is?*
- You should see the aluminum foil ball move away from the edge of the pan. *Why do you think the ball moves like this? Can you explain what is happening?*
- Now, touch the ball with your finger. *What happens?*
- Now that you know your electroscope works, you can use it to test the static electricity present in other materials. To do this, first discharge your electroscope by touching the pan with your finger. Next, rub the material you want to test with the balloon several times to charge the material. Then quickly lift up the electroscope (holding it by its Styrofoam cup) and place the test material on top of the Styrofoam plate so that the material is laying flat on the plate. Make sure the material is not touching the table. Then place the electroscope on top of the object. *What happens to the aluminum foil ball? Why do you think this is?*
- Again, touch the ball with your finger. *What happens this time?*
- *Based on your observations, was the material you tested able to hold a static electric charge?*

EXTRA

Use your homemade electroscope to test even more materials. *Which ones can hold a static charge and which can't?*

SCIENCE FAIR IDEA

You can also use your electroscope to investigate which materials conduct the most static electricity. This is because the farther the aluminum ball moves from the aluminum pan, the bigger the charge from the test material is. *Which common household materials can build the greatest electrical charge; which ones the least?*

SCIENCE FAIR IDEA

Some objects become negatively charged and other objects become positively charged with static electricity. Try to discover a way to investigate this. *Does this kind of electroscope detect both types? How can you tell the difference between the two?*

EXTRA

Static electricity is not good when it gets in your clothes! Try an experiment rubbing an object with a dryer softener sheet (such as Bounce) after rubbing the object against a balloon. *How do dryer sheets work? What happens to the electroscope reading after rubbing a charged object against a dryer sheet? How do different dryer sheet products compare?*

OBSERVATIONS AND RESULTS

Did the aluminum ball on the electroscope move away from the pan when you placed the electroscope on the charged Styrofoam plate? Did the other material you tested behave similarly or not move the ball at all?

When an object, such as the Styrofoam plate, becomes electrically charged, it can be either positive or negative. (If an object has a lot of electrons, it is negatively charged; if it does not have many electrons, it has a positive charge. Whether an object tends to gain or lose electrons depends on the type of material it's made of.) When a charged object (such as the charged Styrofoam plate) touches the aluminum pan of the electroscope, the charge (or electrons) easily moves through the metal pan. Because the aluminum ball touches the pan, the ball gains the same charge as the pan—they are both either positive or negative. Because objects that have the same charge are repelled by one another, the ball is pushed away from the pan. Materials that tend to gain or lose electrons include wool, human hair, dry skin, silk, nylon, tissue paper, plastic wrap, and polyester—and when testing these materials, you should have found that they moved the aluminum ball similarly to how the Styrofoam plate did.

CLEANUP

Return all supplies to where you found them. Recycle or throw away any items that are no longer useful, such as the Styrofoam cup.

Attraction with Static Electricity

LEARN HOW WOOL AND BALLOONS CAN COME TOGETHER TO CREATE STATIC ELECTRICITY.

PROJECT TIME

30 minutes

KEY CONCEPTS

Electricity
Electronics
Insulators
Conductors

Have you ever wondered why rubbing a balloon or a blanket—or even a winter hat—on your head makes your hair stand up? The effect is due to static electricity, but how is the static electricity made, and why does it make your hair stand on end? Static electricity is the buildup of electrical charge in an object. Sometimes static electricity can suddenly discharge, such as when a bolt of lightning flashes through the sky. Other times, static electricity can cause objects to cling to one another. Think of how socks fresh out of the dryer stick together. This happens when objects have opposite charges, positive and negative, which attract. (Objects with the same charges repel one another.) Could enough static electricity make a balloon stick to a wall? How much do you think you would have to rub it?

BACKGROUND

When one object is rubbed against another, static electricity can be created. This is because the rubbing creates a negative charge that is carried by electrons. The electrons can build up to produce static electricity. For example, when you shuffle your feet across a carpet, you are creating many surface contacts between your feet and the carpet, allowing electrons to transfer to you, thereby building up a static charge on your skin. When you touch another person or an object, you can suddenly discharge the static as an electrical shock.

Similarly, when you rub a balloon on your head, it causes opposite static charges to build up both on your hair and the balloon. Consequently, when you pull the balloon slowly away from your head, you can see these two opposite static charges attracting one another and making your hair stand up.

MATERIALS

- Balloon
- An object made out of wool (such as a sweater, scarf, blanket, or ball of yarn)
- Stopwatch
- A wall
- A partner (optional)

PREPARATION

- Blow up the balloon and tie off the end.
- Have your partner prepare to use the stopwatch.

PROCEDURE

- Hold the balloon in a way that your hand covers as little of its surface area as possible, such as by using only your thumb and pointer finger or by gripping the balloon by its neck where it is tied off.

- Rub the balloon on the woolly object once, in one direction.
- Hold the balloon up on the wall with the side that was rubbed against the wool facing the wall, then release it. *Does the balloon stay stuck on the wall?* If the balloon stays stuck, have your partner immediately start the stopwatch to time how long the balloon remains bound to the wall. If the balloon does not stick, move to the next step.
- Touch the balloon to a metal object. *Why do you think this is important to do?*
- Repeat the above process, but each time increase the number of times you rub the balloon on the woolly object. Rub the balloon in the same direction each time. (Do not rub the balloon back and forth.) *How many rubs does it take to make the balloon stick to the wall for a few seconds? What about multiple minutes?*
- You can repeat this whole process two more times. *Do your observations for each trial match with the previous trials?*

EXTRA

Does rubbing in one direction give a different result than rubbing back and forth? Try comparing the same number of rubs in one direction with those done back and forth. *Does one stay on the wall longer than the other?*

SCIENCE FAIR IDEA

Try comparing the effectiveness of different materials for producing a static charge. *Does rubbing wool work better than rubbing silk?* Design an experiment to test several different materials: silk, wool, nylon, polyester, plastic, metal, etcetera.

OBSERVATIONS AND RESULTS

In general, did the balloon stick to the wall for a longer amount of time as you increased the number of times you rubbed the balloon on the woolly object?

Wool is a conductive material, which means it readily gives away its electrons. Consequently, when you rub a balloon on wool, this causes the electrons to move from the wool to the balloon's surface. The rubbed part of the balloon now has a negative charge. Objects made of rubber, such as the balloon, are electrical insulators, meaning that they resist electric charges flowing through them. This is why only part of the balloon may have a negative charge (where the wool rubbed it) and the rest may remain neutral.

When the balloon has been rubbed enough times to gain a sufficient negative charge, it will be attracted to the wall. Although the wall should normally have a neutral charge, the charges within it can rearrange so that a positively charged area attracts the negatively charged balloon. Because the wall is also an electrical insulator, the charge is not immediately discharged. However, because metal is an electrical conductor, when you rub the balloon against metal, the extra electrons in the balloon quickly leave the balloon and move into the metal so the balloon is no longer attracted and does not adhere.

CLEANUP

Deflate and throw out the balloon. Return other objects to where you found them.

Dance with Your Sweater

USE STATIC ELECTRICITY TO MAKE BITS OF PAPER AND RICE JUMP UP AND DOWN!

Have you ever felt a shock when you tap someone on the shoulder or touch a car door? Or have you ever seen sparks when you comb your hair or rubbed a fleece blanket on a cold day? If you have, then you've felt the effects of static electricity! In this activity, you'll learn how to use static electricity to make bits of paper jump up and down.

PROJECT TIME

30 minutes

KEY CONCEPTS

Electric Charge
Polarity
Attraction
Repulsion

BACKGROUND

Everything is made of tiny building blocks called atoms. Although atoms are already microscopic, they contain even smaller particles called electrons, protons, and neutrons. Electrons carry a negative charge and protons carry a positive charge whereas neutrons have no charge. There are usually an equal number of electrons and protons in an atom, and the atom is therefore considered "neutral." But, if an atom gains electrons, it becomes "negatively charged." If it loses electrons, it becomes "positively charged." Atoms with the same charges repel one another and atoms with unlike charges are attracted to one another.

MATERIALS

- Sheet of Plexiglas™ or other clear plastic (about 1 foot [0.3 m] square and 0.25 inch [0.64 cm] thick); the plastic lid from a storage container, large takeout container or poster frame can also work well
- Four supports about 1 to 2 inches (2.5 to 5 cm) high (Tuna cans work nicely.)
- Tiny bits of paper, Styrofoam, puffed rice, plain dry rice grains, or confetti
- Pieces of different types of fabric, including wool
- Flat, smooth table or countertop surface where you can conduct the activity

PREPARATION

- Scatter the paper, Styrofoam, rice, or confetti on a small section of a table.
- Set up the supports—to reach the four corners of your plastic—around the bits of material.
- Place the plastic sheet on the supports. (The small pieces of material should be below the plastic sheet.)

PROCEDURE

- *What are the small bits of material under the plastic doing?*
- "Charge" the plastic by rubbing it vigorously with the piece of wool.
- Watch the bits! *What are they doing? How long did they "dance"?*
- Try charging the plastic by rubbing different types of material on it, including your hand. *What material made the bits dance the longest?*

EXTRA

Remove the bits of paper, Styrofoam, rice, or confetti and scatter a different type of material under the plastic. Repeat the process of charging the plastic. *Are these new bits more active than the previous ones? Which bits were most active? Why?*

OBSERVATIONS AND RESULTS

Both the plastic and the small bits start out electrically neutral. That is, they have a roughly equal number of protons and electrons. When you rub the plastic with the wool cloth, the wool transfers negative charges (electrons) to the plastic, making it negatively charged.

These negative charges polarize the bits, making one side of the paper, Styrofoam, rice, or confetti positive and the other side negative, attracting the positive charges to the tops of the bits and pushing the negative charges away, to the bottoms. The attraction between the negative plastic and the positive charge concentrated on the top of the bits makes them "jump" up to the underside of the plastic.

But, when a bit actually touches the plastic, some of the plastic's negative charge flows to the bit. Because the whole bit of material was originally neutral, the bit now has excess negative charge. The negatively charged bit and the negatively charged plastic now repel each other, which causes the bit to "jump" quickly back to the table.

As the bit's excess negative charge slowly drains away to the tabletop or the air, the bit again becomes neutral and is ready to jump up to the plastic once more, making the group of bits appear to dance. Over time, the negative charge of the plastic dissipates—until you recharge it by rubbing it again!

CLEANUP

Throw out, compost, or recycle the small bits of paper, rice, etc. Return other materials where you found them.

Block Radio Waves

LEARN HOW COMMON HOUSEHOLD OBJECTS CAN COMPLETELY DISABLE A RADIO-CONTROLLED CAR!

PROJECT TIME

30 minutes

KEY CONCEPTS

Physics
Radio waves
Electricity
Electromagnetism
Wireless communication

Have you ever wondered how a radio can receive music and news broadcasts over thin air? Radios, as well as radio-controlled cars and cell phones, all receive information via invisible waves. Some of these waves are called radio waves. Radio waves are a type of electromagnetic radiation, a kind of energy that also includes visible light. Radio waves are not harmful, but are in fact extremely useful for communicating across long distances. To send information using radio waves, a transmitting antenna sends out a radio wave at a certain frequency (which can tell us the size of the wave), and this is picked up by a receiving antenna. Some materials can block, or interfere with, radio waves. Have you ever noticed that you lose reception on a car radio or cell phone when you drive into a tunnel or an underground parking garage? What materials block the radio waves, and which ones allow the waves to easily pass through?

BACKGROUND

A transmitting antenna, or transmitter, generates and sends out radio waves thanks to a specific electrical current. This is possible because inside transmitters are wires that allow negatively charged particles, called electrons, to flow through them, which makes an electrical current. When current flows within a wire, it generates an electromagnetic field around the wire. This electromagnetic field radiates out of the antenna in all directions, creating invisible radio waves.

When the electromagnetic radio waves hit a receiving antenna, or receiver, such as a radio, it generates a current inside of a wire in the receiver. (This is the reverse process of what happened in the transmitter when the current generated the electromagnetic field in the first place.) The receiver then processes the current back into the transmitted information, which, for a radio, allows you to hear music or other broadcasts. Some materials can block the radio waves that the transmitter generates, which can be tested by seeing whether a receiver can process and respond to information sent by a transmitter.

MATERIALS

- Radio-controlled car and its corresponding remote control (both with working batteries)
- Cotton (such as a piece of clothing made from cotton)
- Aluminum foil
- Plastic wrap
- Wax paper
- Rubber glove
- A smooth, wide open space to test drive your radio-controlled car

PREPARATION

- First check the radio-controlled car and its remote control to make sure they both have fresh batteries. Check that the car runs well on the open space that you will be using.

- Make sure you have enough of each material you want to test (the cotton, aluminum foil, plastic wrap, wax paper, and rubber glove) to completely cover the RC car remote control. Every covering should be loose enough so you can still operate the remote control buttons through it.

PROCEDURE

- Wrap the remote control in cotton. Make sure that the remote control is completely and securely covered so that there are not any openings or holes in the covering, and that it is covered loosely enough so that you can operate the controls.
- Try to operate the radio-controlled car using the cotton-covered remote control. *Does it work? Does the car move at all?*
- Remove the tested material from the remote control. Test to make sure that the car still works when it is not covered with any material.
- Repeat this process using each different material separately. *For which materials does the car still operate, and for which materials does the car not operate? Why do you think this is?*

EXTRA

Are the wireless signals transmitted by other devices blocked by the same material(s) that you found could block the signals from the remote control in this activity? You can try this activity again but using other remote control devices, such as your TV or stereo remote. *How do different wireless devices respond? Do you think they use similar or different types of waves?*

EXTRA

Do you need to completely cover the radio-controlled car remote control to block its signal? Using the material(s) you found to block the signals from the remote control to the radio-controlled car, completely cover the remote control as you originally did in this activity, and then slowly remove the covering from the remote control while trying to operate the car. *How much covering is needed to block the signal? Are some parts of the remote control more important to cover in order to block the signal?*

OBSERVATIONS AND RESULTS

Did covering the radio-controlled car remote control with aluminum foil stop the radio-controlled car from working? Did any of the other materials tested stop the car?

When you operate the car using its remote control, the remote control transmits radio waves at specific frequencies that can be received by the car. The remote control acts as a transmitter and the car as a receiver. Then, when the car receives the radio waves, which are electromagnetic waves, the waves generate a current in a wire in the car, and this tells the car which direction to move in. Depending on a material's thickness and composition, it might block—or interfere with—radio waves. Thin amounts of plastic wrap, wax paper, cotton, and rubber are not likely to interfere with radio waves. However, aluminum foil, and other electrically conductive metals such as copper, can reflect and absorb the radio waves and consequently interfere with their transmission. Placing the transmitter or receiver in a fully enclosed container made of highly conductive metal, such as was done in this activity, is the most efficient way to interfere with radio waves.

CLEANUP

Return all materials to where you found them.

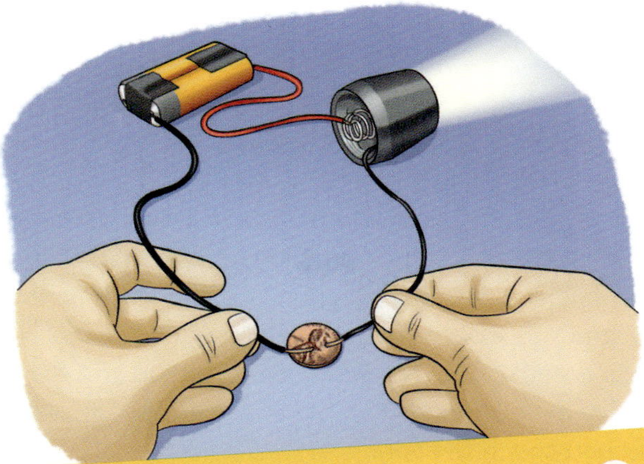

Which Materials Conduct Electricity?

MAKE YOUR OWN FLASHLIGHT—AND SEE IF YOU CAN MAKE THE BULB SHINE BRIGHTLY!

Electricity powers many of the devices you use every day. Those devices are made up of circuits, ranging from very simple (such as a lamp with a single lightbulb) to very complex (such as in a computer). Try this project to build your own simple circuit and use it to test which common household materials conduct electricity.

PROJECT TIME

45 to 60 minutes

KEY CONCEPTS

Electricity
Conductor
Insulator

BACKGROUND

You probably hear the word electricity a lot, but what does it actually mean? In everyday use, electricity typically refers to electrically charged particles (called electrons) moving through metal wires. The flow of electricity is called current. Metals are generally very good conductors, meaning they let current flow easily. Materials that do not let current flow easily are called insulators. Most nonmetal materials such as plastic, wood, and rubber are insulators. You will notice this if you have ever plugged something into a wall outlet. The prongs on the plug and the wire inside the cord are metal, but they are surrounded by plastic or rubber insulation, so you do not get shocked when you touch the cord!

Electricity requires a complete "loop" for current to flow. This is called a closed circuit. That is why wall outlets have two prongs and batteries have two ends (positive and negative) instead of just one. You connect both of them to a circuit and that creates a complete loop. If the loop is broken at all, it becomes an open circuit, and no current will flow.

In this project, you will build your own simple circuit by disassembling a flashlight (with permission, of course). You will use your circuit as a tester to determine whether household materials are conductors or insulators. When you connect the circuit to a conductor, you will create a closed circuit, and the flashlight bulb will turn on. If you connect the circuit to an insulator, you will still have an open circuit so the bulb will stay off.

MATERIALS

- Flashlight (one you can take apart)
- Batteries for your flashlight
- Three pieces of wire that can be cut and stripped (See Procedure for more information.)
- Ruler with metric measurements
- Electrical tape (and/or rubber bands)
- Scissors or knife (and an adult's help)
- Assortment of metal and nonmetal household materials that can be tested in your circuit

PREPARATION

- To do this project, you will need to salvage three pieces of wire from an old electronic device. You might have a junk drawer full of old cell phone chargers—those will work great. You can also purchase wire at hardware or some craft stores.
- Cut three pieces of wire, each at least 10 centimeters long.
- Have an adult use scissors or a sharp knife to shave about 1 centimeter of insulation off the ends of each wire, exposing the metal inside. (There is also a special tool to do this called wire strippers. An adult can use those if they are available.)
- Disassemble your flashlight. Remove the batteries. If possible, unscrew the "head" (the part that holds the bulb) and remove the on/off switch. Most flashlights can be disassembled easily by hand, but you might need another tool (such as a screwdriver) and/or help from an adult to do this.
- **Caution:** Electricity from wall outlets is very dangerous and can be deadly. Never cut into a wire or open an electronic device while it is plugged into a wall outlet.

PROCEDURE

- Examine the inside of your flashlight and try to trace the circuit. Remember that electricity requires a closed circuit to flow. The circuit in a flashlight usually goes from one end of the battery compartment, via an on/off switch, then through the bulb and back to the other end of the battery compartment. *Can you find the circuit?*
- Your first goal is to use two wires to connect the battery compartment directly to the bulb. This might take some tinkering on your part—not all flashlights are the same. *Is it difficult to create your own new closed circuit?*

- The battery compartment should have positive (+) and negative (-) ends. Use electrical tape to attach one end of a wire to the metal parts on each end of the battery compartment. Make sure to squeeze the wires on firmly so they make good contact. (Tip: If the batteries just slide into the flashlight body—rather than being held in place by clips or springs—use rubber bands to hold them together end to end when you remove them from the flashlight.)

- Now search for two metal contacts on the bulb housing and connect the other ends of the wires to those with electrical tape. Tip: Sometimes the entire inside of the flashlight case is metal, and this serves as one of the contacts. *Were you able to create a circuit and make the bulb light up?*

- If you made the contacts correctly, this should cause the bulb to light up. If the bulb does not light, don't worry! There are a few things you can check:

- You might have an LED flashlight. LED stands for light-emitting diode. An LED is a special type of lightbulb that acts like a one-way valve for electricity. It only lights when its positive (+) and negative (-) sides are connected the right way. Try reversing how the two wires are connected to the battery pack and see if it lights up.

- Another reason you might not be getting light is that your wires might not be making good contact with the metal in the flashlight circuit. Try pinching the contact points with your fingers or use something such as mini clothespins or binder clips to squeeze the connections.

- You should now have a working circuit. Essentially you have removed the battery and bulb from the flashlight case and re-created the circuit using two wires. You can use this circuit to test the conductivity of household materials by adding a third wire.

- Disconnect the wire from one end of the battery pack. This creates an open circuit, and your bulb should turn off.

- Tape one end of your third wire to that end of the battery pack. Your circuit should now have three wires total, two of which have free ends.
- Touch the two free ends of the wires together. This should create a closed circuit again, and your bulb should turn on.
- Test whether materials are conductive by touching them with both free wire ends at the same time.
- *What happens if you touch metal objects, such as paper clips or aluminum foil? If the bulb turns on, does that mean the material is a conductor or insulator?*
- *What happens if you touch nonmetal objects such as wood, plastic, or rubber? Does the bulb turn on or stay off?*

EXTRA

Can you find any nonmetal conductive materials in your home?

OBSERVATIONS AND RESULTS

It may take a bit of work to reverse engineer a flashlight once you have taken it apart. You should be able to get the flashlight to function without its power switch, however, by connecting the battery compartment directly to the bulb using two wires. Adding a third wire allows you to create a "tester." When you touch a metal object with the free wire ends, the bulb should light up just like it usually would. This works because the metal objects are conductors, so they create a closed circuit. When you touch insulating materials such as plastic, rubber, and wood, the circuit remains open, so the bulb stays off because no current can flow.

Nonmetal conductive materials can be difficult to find. A graphite pencil core may work for some flashlights. But graphite has a very high resistance compared to metals, so the bulb may appear very dim or not light up at all.

CLEANUP

Reassemble your flashlight if you need to use it again or keep your homemade conductivity tester!

Generate Electricity with a Lemon Battery

DID YOU KNOW YOU CAN MAKE A BATTERY OUT OF A PIECE OF FRUIT? YOU'LL BE CHARGED UP ON SCIENCE WHEN YOU FEEL THE SUCCESS OF YOUR HOMEMADE ELECTRICITY!

PROJECT TIME

45 to 60 minutes

KEY CONCEPTS

Electricity
Batteries
Electrochemical reaction
Electric conductor

Can you imagine how your life would change if batteries did not exist? If it were not for this handy way to store electrical energy, we would not be able to have all of our portable electronic devices, such as phones, tablets and laptop computers. So many other items—from remote-control cars to flashlights to hearing aids—would also need to be plugged into a wall outlet in order to function. In 1800, Alessandro Volta invented the first battery, and scientists have been hard at work ever since improving previous designs. With all this work put into batteries and all the frustration you might have had coping with dead ones, it might surprise you that you can easily make one out of household materials. Try this activity, and it might just charge your imagination!

BACKGROUND

Batteries are containers that store chemical energy, which can be converted to electrical energy—or what we call electricity. They depend on an electrochemical reaction to do this. The reaction typically occurs between two pieces of metal, called electrodes, and a liquid or paste, called an electrolyte. For a battery to work well, the electrodes must be made up of two different types of materials. This ensures one will react differently than the other with the electrolyte. This difference is what generates electricity. Connect the two electrodes with a material that can transport electricity well (called a conductor) and the chemical reactions fire up; the battery is generating electricity! As you make connections, note that electricity likes to take the path of least resistance. If there are multiple ways to go from one electrode to the other, the electricity will take the path that lets it flow most easily.

Now that you know the essentials of a battery, let's examine some household materials. Aluminum foil is a good conductor—electricity flows easily through it. The human body conducts electricity as well, but not as well as aluminum foil. Electrodes are as common as copper pennies you might have stashed in your piggy bank. As for electrolytes, they are found all over the kitchen; lemon juice is just one example. A simple household battery might be easier to make than you imagined!

MATERIALS

- At least two pennies
- Water
- A few drops of dishwashing soap
- Paper towels
- Aluminum foil (at least 3.5 inches [9 cm] by 24 inches [61 cm])
- Scissors
- Ruler
- At least one lemon (preferably with a thin skin)
- Plate
- Knife (and an adult's help when using it)
- At least two plastic-coated paper clips

PREPARATION

- Wash your pennies in soapy water, then rinse and dry them off with a paper towel. This will remove any dirt sticking to them.

- Carefully cut three aluminum foil rectangles, each 1 inch (2.5 cm) by 8 inches (20 cm).
- Fold each strip in thirds lengthwise to get three sturdy aluminum strips.
- **Note:** In this activity, you will make a very low-voltage battery. The amount of electricity generated by this homemade battery is safe, and you will even be able to test it by touching your finger to it and feeling the weak current. Higher voltages of electricity, however, can be very dangerous and even deadly; you should not experiment with commercial batteries or wall outlets.

PROCEDURE

- Place the lemon on its side on a plate and have an adult carefully use the knife to make a small cut near the middle of the lemon (away from either end). Make the cut about 0.8 inch (2 cm) long and 0.4 inch (1 cm) deep.
- Make a second, similar cut about 0.4 inch (1 cm) away and parallel to the first cut.
- Push a penny in the first cut until only half of it is showing above the lemon skin. Part of the penny should be in contact with the lemon juice because that is what serves as the electrolyte. This copper penny in contact with the lemon juice serves as your first electrode. **Note:** If your lemon has a very thick skin, you might need an adult to carefully cut away some lemon peel. *Why do you think is it important for part of the penny to be in contact with the lemon juice?*
- Slide one of the aluminum strips in the second cut until you are sure part of the aluminum is in contact with the lemon juice. *Can you guess which part of a battery the aluminum strip that sits inside the lemon is? Do you think it is important for the aluminum to be in contact with the lemon juice?*

- You have just made a battery! It has two electrodes made of different metals and an electrolyte separating them. *Do you think this battery is generating electricity or is there still something missing?*

- Your battery can generate electricity, but will only do so when the electrodes are connected with something that conducts electricity. To make a connection attach the second aluminum strip to the part of the penny sticking out of the lemon with a plastic-coated paper clip. Make sure the aluminum touches the penny so electricity can pass between the copper and aluminum. *You used an aluminum strip to create a connection; would you expect a plastic strip to work as well? Do you know why you do not need to create a connection to the second electrode for this particular battery?*

- As soon as the two aluminum strips touch one another, electricity will be produced in the battery and flow through the strips, from one electrode to the other. Because you cannot see the electricity flowing, you can try to feel it. Keep the two strips about 0.4 inch (1 cm) apart and touch your fingertip to them. *Can you feel a tingling, created by a small amount of electricity running from one aluminum strip to the other through your body?*

- For more electrical juice (and a slightly stronger tingling sensation), you can build a second battery, identical to the first. You can choose a different spot on the lemon you just used or use a second lemon to build a second battery. Note that you only need one aluminum strip to build a second battery. To connect the second one to the original, find the aluminum strip of the first battery that serves as electrode. (It has its end inserted in the lemon.) Use a plastic-coated paper clip to attach the other end of this aluminum strip to the penny of the second battery. This connects the aluminum electrode of the first battery to the copper electrode of the second battery.

- Test this set of connected batteries in a similar way as you tested the single battery, bringing the ends of the two aluminum foil strips sticking out of your battery set (those that have a free end) in contact with your fingertip. *Can you feel electricity running? If you could feel it well the first time, is this any different?* (**Note:** If you cannot feel the tingling sensation, check if each electrode—pennies and the aluminum strips stuck in the lemon—are inserted deep enough so they are in contact with lemon juice; make sure there is firm contact between the penny and its attached aluminum strip; and that the aluminum strips are not touching one another. If all is correct, maybe you need slightly more electricity to feel tingling. You can test another person to see if they can feel the electricity, or you can opt to add one more lemon battery to your set.)

EXTRA

Now that you can detect whether electricity is generated or not, try some different configurations. *What happens if you let the aluminum strips touch? What happens if you replace an aluminum strip with a plastic piece, an unfolded metal paper clip, or a toothpick?*

EXTRA

Scientists call the way you connected your batteries in this activity "connecting batteries in series." *Do you think the way you connect two batteries makes a difference in the amount of electricity you felt?* Try it out by connecting the two copper electrodes to one another and attaching the two aluminum electrodes in the same way. (**Note:** You will need an extra strip of aluminum to do this.) Scientists call this "connecting batteries in parallel." Test both ways of connecting batteries and compare. *Do you feel a difference?*

EXTRA

Try different types of metals as electrodes for your batteries. *Do you think a battery with two pennies as electrodes would generate electricity? What about a battery with a penny and a nickel?* Note that some combinations might generate electricity, but the amount generated might be below your ability to feel it. Connecting two or more of these batteries might help you identify good combinations.

SCIENCE FAIR IDEA

You used a lemon to provide the electrolyte for your battery. *Do you think other vegetables or fruits would work as well? Would a potato, apple, or onion battery work?* Try a few from around the kitchen (with permission, of course). *Does one particular fruit or vegetable outperform the others? With what you learned about how batteries generate electricity, why do you think that one type of produce made a stronger battery?*

SCIENCE FAIR IDEA

If you have an LED (light-emitting diode) available, investigate how many lemon batteries are needed to light it.

OBSERVATIONS AND RESULTS

Did you feel the tingling in your fingertip?

The battery you just made has a copper and an aluminum electrode separated by electrolyte lemon juice. It will generate electricity as soon as the electricity has a path to flow from one electrode to the other. You created this path using strips of aluminum, a material that conducts electricity well.

By connecting your battery to your fingertip, you allowed the small amount of electricity it generates to run through your body. This amount of electricity can create a tingling feeling in a fingertip. Experiences will differ from person to person. Some people might only feel the bigger signal generated by connecting several batteries in a particular way. Letting the aluminum strips touch provides a very easy way for the electricity to run from one electrode to the other, so almost no electricity will travel through your body and the tingling sensation disappears. Plastic and wood do not conduct electricity well; none will be felt when using these materials as connections. Metals, on the other hand, conduct electricity well. Different combinations of metals as electrodes will influence the amount of electricity generated. Using identical metals as electrodes will not generate electricity, however.

In this activity, you made a very low-voltage homemade battery. But using commercial batteries can be dangerous—and never experiment with wall outlets!

CLEANUP

Throw away the lemon and aluminum strips, and return other items to where you found them. Wipe up any lemon juice that might have spilled onto your work surface.

Flying Stick Chain Reaction

LEARN HOW TO MAKE STICKS GO FLYING ALL WITHOUT THE USE OF ELECTRICITY.

If you enjoy chain reactions and spectacular sights, this activity is for you! Can you imagine how weaving rather unspectacular tongue depressors (or popsicle sticks or craft sticks) could set off a chain reaction of sticks flying in the air? Try it out and see for yourself!

PROJECT TIME

45 to 60 minutes

KEY CONCEPTS

Physics
Energy
Potential energy
Conservation of energy

BACKGROUND

Energy is never created; it always gets transferred from one kind to another. This activity is a great example of this key physics principle!

If you have ever stretched a rubber band, you have felt how it wants to come back to its original length. Scientists say that a stretched rubber band is loaded with elastic potential energy. In a similar way, a wooden tongue-depressor stick (like the kind used at the doctor's office) stores elastic potential energy when you bend it (though not if you break it). When you allow it to go back to its flat state, it releases this energy.

If a bent depressor stick straightens up near a hard floor, the stick might push itself off of the floor and fly into the air. The elastic potential energy that was stored in the stick is transformed into movement, called kinetic energy. As it flies up, it gains height above the floor, which translates into gravitational potential energy, or the energy due to the fact that it can fall back down. Eventually the stick lays flat on the ground. The energy once stored in the stick has ultimately been transformed into heat and sound—other forms of energy.

When one tongue depressor stick releases its stored energy, nothing too spectacular happens. But what will happen when a chain of sticks release their energy?

MATERIALS

- At least 50 tongue depressors. If these are not available, large, somewhat flexible popsicle sticks or wide craft sticks are a good alternative.
- Hard floor
- Four markers: light blue, dark blue, orange, and red
- Metric ruler
- Glue
- Pen (optional)
- Video camera or device that takes video (optional)

PREPARATION

- Color four tongue depressors light blue, four dark blue, four orange, and seven red.

- Make a piece to start your stick chain by placing one light blue, one dark blue, and another light blue tongue depressor parallel to each other. Leave about 2 centimeters between the depressors.
- Stagger the parallel depressors so each depressor sticks out 2 to 3 centimeters above the adjacent one.
- Place a red depressor at a right angle to all three depressors. Place it on top of the three depressors, and adjust it until it just touches all three of them. Glue it in place.
- Place a second red depressor next to the first red one, and glue it to all three staggered depressors. Let the glue dry.
- Turn the structure over, and glue two red depressors matching the ones of the other side. Now the three blue staggered depressors will be sandwiched between the red ones. Allow the glue to dry.
- The light blue depressor that sticks out most will be "1" (you can use a pen to label it if you like). The middle (dark blue) depressor will be "2," and the one that sticks out the least (light blue) will be "3."

PROCEDURE

- To start your chain, gather the starter piece that you assembled (ensuring that the glue is dry), the few colored tongue depressors that are left and a large number of uncolored tongue depressors on your hard floor.
- Slip an orange depressor in the starter piece so it goes under depressor 1 (light blue), over depressor 2 (dark blue) and under depressor 3 (light blue). It should stick out from under depressor 3 just a little. Shift it until it is parallel to the red depressors from the starter piece, leaving about 2 centimeters between them.

- Slip a red depressor in the starter piece so it goes over depressor 1 (light blue), under depressor 2 (dark blue) and sticks out from under depressor 2 just a little. Shift it until it is parallel to the orange depressor, leaving about 2 centimeters between them.

- You should see two parallel depressors (the orange and the red depressors you just added) sticking out. Slip a dark blue depressor under the shorter one (the orange one) until it sticks out a little and over the longer one (the red one), parallel to depressor 1 (light blue). You might feel you have to hold the depressors in place. *Why would this be?*

- You should now see two parallel depressors sticking out: a light blue and a dark blue one. Slip an orange depressor under the shorter one (light blue) until it sticks out a little and over the longer one (dark blue) so it makes a right angle with the sticks it crosses.

- Find the two parallel depressors sticking out: they are red and orange. Slip a light blue depressor under the shorter one until it sticks out a little and over the longer one so it makes a right angle with the depressors it crosses. *Do you notice that all the blue depressors are parallel, and that all the red and orange depressors are parallel?*

- Find the two parallel depressors sticking out: they will be at a right angle with the last depressor you added. Slip a depressor under the shorter of the two until it sticks out a little and over the longer one so it makes a right angle with the depressors it crosses.

- Repeat the previous step multiple times. *Do you feel like you have to bend the depressors? How would that help create an energetic chain reaction? What do you think will happen when you no longer hold down the last couple of depressors you weaved into the chain?*

- Keep building until you are satisfied with the size of your chain. You can keep the depressors in place by fixing the last depressor you are adding at a diagonal under the one that would be parallel to it if you were continuing making the chain longer.
- When you are ready to let your chain reaction go, watch closely, and let the last depressor go. *What happens? Why would this happen? Did you notice other forms of energy as the reaction occurs?*

EXTRA

Build a longer chain and see how the chain reaction changes as the chain gets longer.

EXTRA

If you have a video camera (or phone or other device that takes video) available, take a video of the chain going off, preferably in slow motion. *What can you learn from the slow-motion video?*

EXTRA

For a challenge, make your chain bend, or make it go over obstacles.

SCIENCE FAIR IDEA

Compare the chain reaction on carpet to a similar chain on a hard floor. *Which one makes the sticks jump up highest? Why would this be the case?*

EXTRA

Make chains from extra-large craft sticks and tongue depressors. *Can you explain why the result is different?*

OBSERVATIONS AND RESULTS

You probably saw the tongue depressor sticks jump up in the air as soon as you released the last one—and maybe even before you planned to let them go.

You might have noticed that you had to bend the depressors in order to make them fit into the chain. With each bend you added energy to the system. Did you feel how the chain was under tension and how you had to hold the last couple of depressors to keep them in place? When you let it go, all this tension is released in a chain reaction. It starts with the last depressors you added straightening. This releases the next depressors in the chain and so on. Soon all the depressor sticks fly in the air, releasing the energy stored in them—and you have a whirlwind of sticks flying around!

If you assembled your chain and left it out for a day or so, you might have noticed your depressors did not jump as high when you released the chain. Over time the depressors yield under the tension and become permanently bent. Part of the energy goes into bending the material (usually wood).

If you tried to release the chain on carpet or any other soft ground, you might also have noticed the depressors did not jump as high. The depressors cannot bounce off the soft ground as easily as they can off of hard flooring because the softer ground absorbs part of their energy.

Maybe you noticed it is harder to weave shorter or more rigid popsicle sticks together. These sticks do not bend as easily; instead they tend to break. They are not as well suited to store elastic potential energy.

CLEANUP

Gather up your materials and return them to where you found them. Save the tongue depressors you used for another project!

The Colossal Cookie-Tin Race

DO ALL OBJECTS WITH THE SAME MASS AND SHAPE ACCELERATE AT THE SAME SPEED?

It is fun to race, especially when the competition is evenly matched! In this race, the force of gravity provides the power, and the competitors are the same size, weight, and distance from the finish line. Gather the materials and see who the winner will be!

PROJECT TIME

30 minutes

KEY CONCEPTS

Physics
Mass
Potential energy
Kinetic energy

BACKGROUND

If you roll two soccer balls down a hill, they should roll similarly—and take about the same amount of time to complete the path. But what if you had two objects that were the same size and the same weight—but had their weight distributed differently? Two objects like this may behave differently when they roll downhill. How quickly an object accelerates depends partly on how its mass is distributed. How? Gather some basic materials and set up a race!

MATERIALS

- Two identical round metal cookie tins (the type butter cookies come in)
- 10 large metal washers, about 0.25 pound (113 g) each
- Double-sided foam tape
- Sidewalk chalk
- Hilly street with a smooth, clear sidewalk

PREPARATION

- Using double-sided foam tape as an adhesive, arrange five of the washers evenly around the rim of the bottom of one metal cookie tin.
- Use double-sided foam tape again to secure the other five washers to the middle of the bottom of the second container.
- Gather all materials and head outside. Find a hill with a smooth sidewalk clear of cars, pedestrians, and obstacles.

PROCEDURE

- Use chalk to mark a starting line at the top of the hill.

- Use chalk to mark a finish line at the bottom of the hill.
- Place both tins at the starting point.
- *Do the tins feel the same in your hand? Which tin do you think will reach the finish line first? Why?*
- Release both tins at the same time so that they can race down the hill. *What do their rolling patterns look like?*
- *Did they reach the finish line at the same time? Was one tin faster than the other?*
- Try racing the tins down the hill again. *Did you get the same results?*

EXTRA

It is also interesting to experiment with rolling cans of soup down an incline. Get a can of thin broth soup and a can of thick cream soup and roll them down the hill. *Did one type of soup consistently win? Why might this be?*

OBSERVATIONS AND RESULTS

The results of the race have to do with potential versus kinetic energy. At the start of the race, both cookie tins have identical potential energy because each has the same mass and the same dimensions. But because the washers are taped in different places (the outside of one tin versus the center of the other), each tin has its mass distributed differently. This affects how quickly the object can accelerate.

The tin with its mass at the rim will use a greater part of its original potential energy just to get rolling. The tin with its mass concentrated at the center has an easier time getting started rolling and thereby doesn't use as much of its potential energy. This means the tin with its mass at the rim has less energy available to translate into kinetic energy, resulting in a lower linear speed.

If you tried racing the cans of soup, a similar difference in kinetic energy occurs. The thin, liquid broth stays relatively in place and does not rotate when the can does. This means the potential energy of the broth can go into linear motion, not into rotating the can. The thick, creamy soup is like a solid inside of the can and rotates with it. This means some of the potential energy of the thick soup is needed for rotational energy; this makes the can roll slower.

CLEANUP

Return materials where you found them. Throw away the used tape.

Strong Wind Science
The Power of a Pinwheel

EXPLORE HOW THE WIND CAN BE TRANSFORMED INTO ENERGY YOU CAN USE!

Have you ever ridden your bike into a strong wind? If so, did it feel really tough? How does this compare with how you feel when the wind is pushing against your back? Does that make you feel ready for the Tour de France? In this science activity, you will explore how wind-powered devices, such as wind turbines and pinwheels, also react in different ways to wind direction.

PROJECT TIME

15 to 45 minutes

KEY CONCEPTS

Energy
Power
Forces
Machines

BACKGROUND

Wind turbines are machines that change the energy in wind into mechanical or electrical energy. Windmills are examples of wind turbines that convert wind energy into mechanical energy. The Netherlands is a country well known for its windmills that have been used for centuries to grind corn, drain land, and cut wood. Wind farms, on the other hand, are examples of wind turbines that convert wind energy into electrical energy. In California, you can see rows of wind turbines along windy ridges and mountain passes. The wind turbines on these wind farms connect directly into power grids and produce 5 percent of the electricity that the entire state uses.

A wind turbine has a rotor with blades that is connected to a shaft. As wind energy hits the blades, the rotor turns, which causes the shaft to turn as well. As the shaft turns, it is able to do work and produce either mechanical or electrical energy. The more wind that hits the blades, the more the rotor can turn, and the more energy the wind turbine can produce.

MATERIALS

- Pinwheel (If you do not have a pinwheel, you can make a homemade one using a sheet of paper, a pencil or pen, a round wooden skewer, and scissors—see steps below.)
- Hair dryer (optional)

PREPARATION

- If you do not have a pinwheel, make one now by going through the following steps.
- Cut a sheet of paper so that it is a square (8.5 inches [21.6 cm] by 8.5 inches [21.6 cm] is ideal).

- Fold the square of paper along the diagonal, then unfold it back into a square again. Fold it along the other diagonal, and then unfold it once again. Your paper should now look like a square with a big "X" made by the creases. About 2 inches (5 cm) out from the center, along each crease, make a small mark with a pen or pencil.
- Carefully make four holes with a skewer to the right of each crease near the square's corners and a hole in the paper's center.
- From each corner of the square, cut down along the crease until you reach the mark you made.
- Lastly, pick up each of the holes at the corners and fold them, one at a time, over onto the skewer, so that they are all on top of one another. You should now have a functional, homemade pinwheel!

PROCEDURE

- If you want, you can use a hair dryer to blow on the blades instead of blowing on them yourself. If you do this, keep the hair dryer on a low setting and be sure it is about the same distance from the blades each time you blow on them.
- Face the front of your pinwheel and blow directly at it. Imagine that the pinwheel is a wind turbine, and when it spins in this direction it is turning the energy of the wind into electrical energy. *Which way do the blades spin, clockwise or counterclockwise? How fast do they turn?*
- Turn the pinwheel so that its front side is now pointing to your right. You should now be facing the side of the blades that were previously on your left. Blow on the top half of the blades, above the shaft that goes through the blades. Try to blow similarly (with a similar force and distance) as you did before. *Which way do the blades spin, clockwise or counterclockwise? How fast do they turn? How does it compare with blowing on the front of the pinwheel?*

- Still facing this side of the blades, blow on the bottom half of the blades, below the shaft. Try to blow similarly to the way you did before. *Which way, and how quickly, do the blades spin now? How does this compare with blowing at the pinwheel from the other directions?*
- Now turn the pinwheel so that its front side is pointing to your left and you are facing the opposite side of the blades. Blow on the top half of the blades, similarly to how you blew on them before. Then blow on the bottom half of the blades. *Which way and how quickly do the blades spin each time?*
- *Overall, how does the direction of the wind hitting the pinwheel affect how the blades spin? If the pinwheel were a wind turbine and blowing on the front of it generated electrical energy, which direction(s) should it receive wind from to efficiently make electrical energy?*

EXTRA

Try repeating this activity with a few different pinwheels. *Do you get the same results for each one?*

SCIENCE FAIR IDEA

If you want a more challenging activity, you can try to quantify how much power the pinwheel makes when it is blown on from one direction compared with another direction. To do this, you will need to attach the pinwheel blades to a fixed shaft (for example, a wooden skewer) so that they rotate the shaft when they are blown on. Support the shaft by putting it through a cardboard oatmeal canister or similar object. On the other end of the shaft, tie a piece of thread with small weights (for example, paper clips) on it. When the pinwheel is blown on and the shaft turns, it should pull up the weights.

OBSERVATIONS AND RESULTS

When you blew straight onto the front of the pinwheel did it spin counterclockwise? Did it spin best when you were blowing into its "cups"?

Most pinwheels have the blades arranged so that when wind blows straight at them, they spin counterclockwise. This is because the blades' "cups" are made so that the oncoming air is captured and pushes the blades in this direction. (You can try looking at several different pinwheels to see this pattern shared among them.) When wind is blown into the cups, the pinwheel spins well. Consequently, if you are using a typical pinwheel, turn it so that the front side is facing your right and blow into the side of the blades. The blades will spin quickly counterclockwise if you blow on the bottom half (into the cups) but will spin slowly clockwise if you blow on the top half (against the backs of the cups). Similarly, if the pinwheel's front side is facing your left and you blow into the side of the blades, they will spin quickly counterclockwise if you blow on the top half (into the cups) but will spin slowly clockwise if you blow on the bottom half (against the cups' backs). If your pinwheel were a wind turbine, and spinning counterclockwise turned wind energy into electrical energy, then it would most efficiently produce electricity when wind blows directly into the cups.

CLEANUP

Return the materials you used to where you found them. You can save your pinwheel for another project!

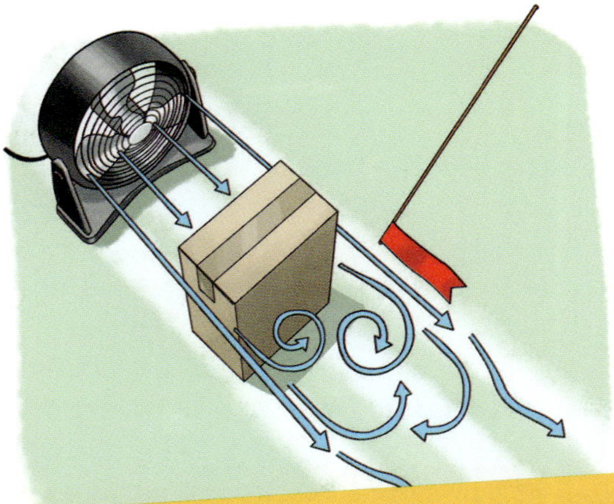

Wild Winds

Turbulent Flow Around Structures

ENERGY-EFFICIENT: WHAT SORT OF BREEZE MAKES WIND TURBINES SPIN BEST? LEARN HOW ENGINEERS PICK THE BEST PLACES TO GENERATE WIND ENERGY IN THIS FUN SCIENCE ACTIVITY.

PROJECT TIME

15 to 45 minutes

KEY CONCEPTS

Physics
Wind power
Aerodynamics

Have you ever seen a wind turbine? Wind turbines are large towers with blades on top that are spun by the wind. They are one source of clean, renewable energy. They use the movement from the wind-spun blades to generate electricity. But in order for the turbines to work best, environmental engineers have to figure out where to put wind turbines so that they get the best exposure to steady, consistent winds. In this project, you will play the role of an environmental engineer and figure out whether it is a good idea to put wind turbines near tall buildings.

BACKGROUND

Have you ever watched water flow through a stream or river? If there are no obstructions in the way, the water probably flows in a smooth, straight line. This is called laminar flow. If there is something in the water, such as a rock or a branch, however, this causes the water to swirl and tumble as it flows by. This is called turbulent flow. The individual "swirls" are called vortices (the plural of vortex).

The same thing happens in air when wind blows around a tall structure, such as a building or a tree. This is an important factor to consider when placing wind turbines, because they will generate the most power from steady winds that blow in the same direction. Inconsistent, turbulent flow can decrease power output or even damage the turbines. In this project, you will use cardboard boxes to represent "buildings" and see how they affect the flow of "wind" blowing from a fan.

Of course, turbulent flow in air is much harder for us to see than a turbulent stream! Sometimes scientists do controlled experiments using colored smoke to observe airflow. In this project, you will use a streamer tied to the end of a stick to visualize the airflow.

MATERIALS

- Fan
- Stick or rod, a few feet long
- Streamer, a few inches long
- Cardboard box (The size of the box depends on the size of your fan—the larger the fan, the larger the box. If the box is empty, you might need to put something inside it to weigh it down.)

PREPARATION

- Set up the cardboard box a few feet from the fan, with the fan pointed at the box.
- Tie the streamer to the end of the stick.

PROCEDURE

- Turn the fan on.
- Standing to the side so your body does not interfere with the airflow, use the stick to hold the streamer "upwind" of the box (on the side of the box facing the fan). *Which direction does the streamer blow? Is its motion consistent? Do you think this shows turbulent or laminar flow?*
- Now move the streamer to the "downwind" side of the box (on the side of the box facing away from the fan) but hold it fairly close to the box. *What happens to the streamer now? Is the flow in this area turbulent or laminar?*
- Gradually move the streamer farther and farther away from the box on the downwind side. *What happens as you move the streamer farther away from the box? What does that suggest about the airflow?*
- *What conclusions can you draw about the airflow near tall structures? How would this affect the placement of wind turbines?*

EXTRA

See how your results change if you change the speed of the fan.

EXTRA

Try the same activity with different size boxes. *How do the results change?*

EXTRA

Try the activity with different or irregularly shaped objects, such as furniture or houseplants. *How do the results change?*

EXTRA

Take your stick and streamer outside on a windy day. Determine which way the wind is blowing and find large objects such as buildings or trees to test the airflow. *What happens if you hold the streamer near the corner of a building or behind a tree?*

OBSERVATIONS AND RESULTS

When you held the streamer "upwind" of the box, you should have seen it blow consistently in one direction, away from the fan. This is because the flow upwind of the structure is laminar (or straight). When you held the streamer just "downwind" of the box, however, it might have fluttered around in all directions or even blown backward. This is because the structure interrupts the smooth flow and causes turbulence. As you gradually moved the streamer away on the downwind side, its motion should have again become consistent, blowing away from the fan. This is because the flow gradually "recovers" and becomes laminar again after the turbulence is introduced.

This is important when placing wind turbines because they should not be too close to large buildings on the "downwind" side. The buildings will create turbulent flow, which will decrease the performance of the wind turbines.

CLEANUP

Return the objects you used to where you found them.

THE SCIENTIFIC METHOD

The scientific method helps scientists—and students—gather facts to prove whether an idea is true. Using this method, scientists come up with ideas and then test those ideas by observing facts and drawing conclusions. You can use the scientific method to develop and test your own ideas!

Question: What do you want to learn? What problem needs to be solved? Be as specific as possible.
Research: Learn more about your topic and refine your question.
Hypothesis: Form an educated guess about what you think will answer your question. This allows you to make a prediction you can test.
Experiment: Create a test to learn if your hypothesis is correct. Limit the number of variables, or elements of the experiment that could change.
Analysis: Record your observations about the progress and results of your experiment. Then analyze your data to understand what it means.
Conclusion: Review all your data. Did the results of the experiment match the prediction? If so, your hypothesis was correct. If not, your hypothesis may need to be changed.

GLOSSARY

accelerate: To increase in speed.

dissipate: To break up and scatter.

distribute: To give or deliver.

efficient: Able to give the wanted effects without waste.

electrolyte: A nonmetallic conductor of energy.

graphite: A soft, black form of carbon often used in lead pencils.

kinetic: Relating to motion and energy.

linear: Relating to a straight line.

neutral: Having no charge.

power grid: A network of electrical lines that provide energy to a geographical area.

radiate: To send out from a center point.

repel: To push away.

rotor: The part of an electrical machine that rotates.

susceptible: Being open to some influence.

vortex: A spinning mass of air or water.

ADDITIONAL RESOURCES

Books

Hardyman, Robyn. *Geothermal Energy*. Bridgnorth, UK: Cheriton Children's Books, 2022.

MacCarald, Clara. *Energy from the Sun*. Lake Elmo, MN: Focus Readers, 2022.

Van Fleet, Carmella. *Electricity: Circuits, Static, and Electromagnets*. Norwich, VT: Nomad Press, 2022.

Websites

Discovery Education
sciencefaircentral.com

Exploratorium
www.exploratorium.edu/search/science%20fair%20projects

Science Buddies
www.sciencebuddies.org/science-fair-projects/project-ideas/list

Science Fun for Everyone!
www.sciencefun.org/?s=science+fair

Videos

Can You Cook a S'more Without a Fire or Electricity?
www.pbslearningmedia.org/resource/solar-oven-smores/can-you-cook-a-smore-without-a-fire-or-electricity/

Pendulum Art
www.pbslearningmedia.org/resource/353132dc-6290-4880-90b6-a42963a736d1/pendulum-art-diy-smarter/

Thermal Energy
www.pbslearningmedia.org/resource/acf101a1-d8e0-44d1-81e3-f427c3c66751/thermal-energy-video-diy-science-time/

Publisher's note to educators and parents: Our editors have carefully reviewed these websites to ensure that they are suitable for students. Many websites change frequently, however, and we cannot guarantee that a site's future contents will continue to meet our high standards of quality and educational value. Be advised that students should be closely supervised whenever they access the internet.

INDEX